Science See-Throughs That Teach

by Ivy Rutzky

NEW YORK · TORONTO · LONDON · AUCKLAND · SYDNEY

To Jacqueline and Julius Rutzky

Thanks to Liza Charlesworth, Deborah Schecter, Patricia J. Wynne,
Donald Silver, Ed Krupp, John G. Maisey,
and the many others who were gracious enough to help.

Cover design by Vincent Ceci and Jaime Lucero
Cover art by Ivy Rutzky
Interior design by Béatrice Schafroth
Cover photographs by Donnelly Marks
Interior illustrations by Ivy Rutzky

ISBN: 0-590-60343-4

Printed in the U.S.A.

Contents

Introduction

What's inside? Whether the object is a child's own body or a faraway volcano, kids want to know! Children are naturally curious about the parts of our world that are hidden from their view. This book was created to encourage that curiosity.

The 10 topics included in this book, both ordinary and extraordinary, were chosen to delight your students and fit with your curriculum. There is a See-Through that reveals each topic's "inner secrets." Giving students a see-through view of an object or environment is more than just a motivational tool. It's also a terrific way to learn about the topic literally "inside and out." With each See-Through, you'll also find a cross-curricular extension that encourages students to search beyond what they can see.

I hope this book will fill your class with lots of "A-hahs!"

—Ivy Rutzky

About This Book

The 10 topics in this book cover animals, plants, habitats, the human body, Earth, and space. Here's an overview of what you'll find:

- **Science See-Through** Double-sided reproducibles that give students an inside look at an interesting object or environment. With these pages, students will be able to compare babies inside four different animals' eggs, or trace the flow of magma up through a volcano.
- **Using the See-Through** Discussion questions to help you guide students to understand what they're viewing—for example, how the structures they see inside a tulip attract bees and help it develop seeds.
- **Background** This section explains important concepts, defines additional vocabulary, and adds other interesting facts to share with students.
- **Coloring Tips** Suggestions to help students color their See-Throughs accurately. Water-based markers or crayons, shaded lightly, work well.
- **Extending Learning** Reproducible activities that connect to other curriculum areas—for example, students use math and graph-reading skills to compare dinosaur lengths. And after learning about constellation myths from different countries, students sharpen their language arts skills by writing an original myth about the stars we call the Big Dipper.
- **Book Links** Suggested fiction and nonfiction books that connect with each of the topics in this book are listed on page 7.

Tips for Photocopying

- Make an equal number of copies of the front and back of a See-Through. Give each student a copy of both. Direct students to the triangles at the top and bottom of the pages. Then show students how to stack the pages back to back. Have students hold the sheets up to the light and line up the triangles. As they hold the sheets in place, you or a student helper should travel around the class stapling or taping the pages together. For best results, staple them twice—once at the bottom and once at the top. Use scissors to trim off uneven edges.

Introducing the See-Throughs

Place a copy of the See-Through faceup on each student's desk. (You may want to do this before students arrive or while they are out of the room—at lunch, for example.) To pique their curiosity, ask that they leave the page faceup until you tell them otherwise.

Use the question at the top of the See-Through and the discussion questions on the teacher's page to get the class thinking. Then hold your page (with the front facing you) up to a light (a lamp or unshaded window) and direct students to do the same. Allow time for students to experiment with different ways to view the see-through effect (front to back, back to front, using different sources of light, etc.). Then return to the discussion questions to help students explore the See-Through view. When appropriate, the discussion questions will direct you to the back of the See-Through as well, where students can better see labeled parts.

Extending Learning

- Science See-Throughs are an excellent way to launch a new science unit. You'll also find a reproducible Extending Learning activity for each topic. Many of these extensions can be assigned as homework.
- Challenge students to research related topics and create their own See-Throughs. Here are some possibilities:

Shark & Whale: fish, seals, other sea life

Eggs: other examples of eggs

Tyrannosaurus rex: other kinds of dinosaurs

Tulip: other kinds of flowers, seeds, trees, mushrooms, etc.

Hidden Homes: beaver lodges, prairie dog towns, etc.

Cave: a cave with different formations and cave animals

Human Skeleton: a close-up view of a hand or foot

Healthy & Decayed Tooth: the process of digestion inside the body

Volcano: rock layers inside a nonvolcanic mountain

Constellations: other constellations

In addition, students may want to present their information in other ways, such as with models, hands-on science activities, or written reports. The Book Links on the next page include valuable resources that students can use for general research or as reference for making their own See-Throughs. The list also includes some excellent fiction entries.

Book Links

Shark & Whale
- *Questions and Answers About Sharks* by Ann McGovern (Scholastic, 1995)
- *Winter Whale* by Joanne Ryder (Morrow Junior Books, 1991)

Eggs
- *Chickens Aren't the Only Ones* by Ruth Heller (Sandcastle Books, 1981)
- *Egg Story* by Anca Hariton (Dutton, 1992)

Tyrannosaurus rex
- *Digging up Tyrannosaurus rex* by John R. Horner and Don Lessum (Crown, 1992)
- *101 Questions About Dinosaurs* by Philip J. Curie and Eva B. Koppelhus (Dover, 1996)

Tulip
- *The Reason for a Flower* by Ruth Heller (Grosset & Dunlap, 1983)
- *National Audubon Society First Field Guide: Wildflowers* (Scholastic, 1998)

Hidden Homes
- *Animals Underground* by Charlotte Ruffault (Young Discovery Library, 1988)
- *Who Lives Here?* by Maggie Silver (Sierra Club Books for Children, 1995)

Cave
- *One Small Square: Cave* by Donald M. Silver and Patricia J. Wynne (McGraw-Hill, 1997)
- *Zipping, Zapping, Zooming Bats* by Ann Earle (HarperCollins, 1995)

Human Skeleton
- *The Glow-in-the-Dark Book of Animal Skeletons* by Regina Kahney (Random House, 1992) and *The Glow-in-the-Dark Book of Human Skeletons* by Michael Novak (Random House, 1997)
- *3-D Kid* designed and engineered by Roger Culbertson (Freeman, 1996)

Healthy & Decayed Tooth
- *Arthur's Tooth* by Marc Brown (Little, Brown, 1985)
- *Dragon Tooth* by Cathryn Falwell (Clarion, 1996)

Volcano
- *The Magic School Bus inside the Earth* by Joanna Cole (Scholastic, 1987)
- *Volcanoes* by Seymour Simon (Mulberry, 1988)

Constellations
- *The Big Dipper and You* by E.C. Krupp (Morrow, 1989)
- *Follow the Drinking Gourd* by Jeanette Winter (Knopf, 1988)

Shark & Whale

Using the See-Through (pages 9–10)

Ask students: How are this great white shark and the minke (MEEN-kuh) whale similar? (They both live in the ocean, swim, and have fins and a submarinelike shape.) Use the points listed below to discuss how they are different.

◆ The whale is bigger than the shark.

◆ The shark has sharp teeth—5 rows in all! It uses them to bite and tear its prey, but the shark doesn't chew up its food before swallowing. Instead of teeth, the whale uses fringed plates that look like big combs, called *baleen*, to filter small fish, plankton, and krill (tiny crustaceans) from sea water. (Only 10 whale species are baleen feeders. Most species have teeth.) This whale also has *throat pleats* that expand as it gulps in huge amounts of water and food.

◆ Some students may know that the shark is a fish while the whale is a mammal. That means, among other things, that the shark uses its *gill slits* to take in sea water which contains oxygen. A whale breathes through its *blowhole* in the same way we use our noses. Also, a whale is warm-blooded (it maintains a body temperature separate from that of its environment) while a shark is cold-blooded (its body temperature changes with its environment). Finally, whales nurse and care for their young; sharks do not.

◆ The tips of the shark's tail point up and down; the whale's tail goes from side to side.

Ask: What differences do you see now?

◆ A shark has bones in its fins and tail; a whale has no bones in its tail or dorsal fin.

◆ The shark has *gills* that help it absorb oxygen from sea water; the whale has *lungs* that absorb oxygen.

Background

Minke whales average 26 to 30 feet long. Great white sharks are commonly 12 to 15 feet, but range up to about 20 feet. (The See-Through shows one of these super-sized specimens.)

Explain to students that the shark is a superior hunter. Most of its brain is devoted to seeing and smelling. A shark can smell with its entire body, not just its nose! When a shark hunts, it first smells its prey, for example, a wounded fish. The shark also senses movements in the water, such as those caused by a struggling fish. Finally, the shark zeros in on its prey by sensing the animal's electrical field, even in dark, murky water.

The minke whale is dark bluish-gray on top of its body and lighter gray below. Its flippers usually have large white patches. Despite its name, the great white shark is a light bluish-gray on top of its body and a dusky white below.

How Are Sharks & Whales Alike? (page 11)

Students use a Venn diagram to compare whale and shark attributes.

Materials: scissors, glue, and a copy of page 11 for each student

Answers: Sharks—can sense electricity, are fish, do not care for their babies, are cold-blooded; whales—breathe air with lungs, are mammals, nurse their babies, are warm-blooded; both—live in the ocean, have fins

Shark & Whale

How are these two giants of the sea alike and different?

Hold this page up to the light to find out!

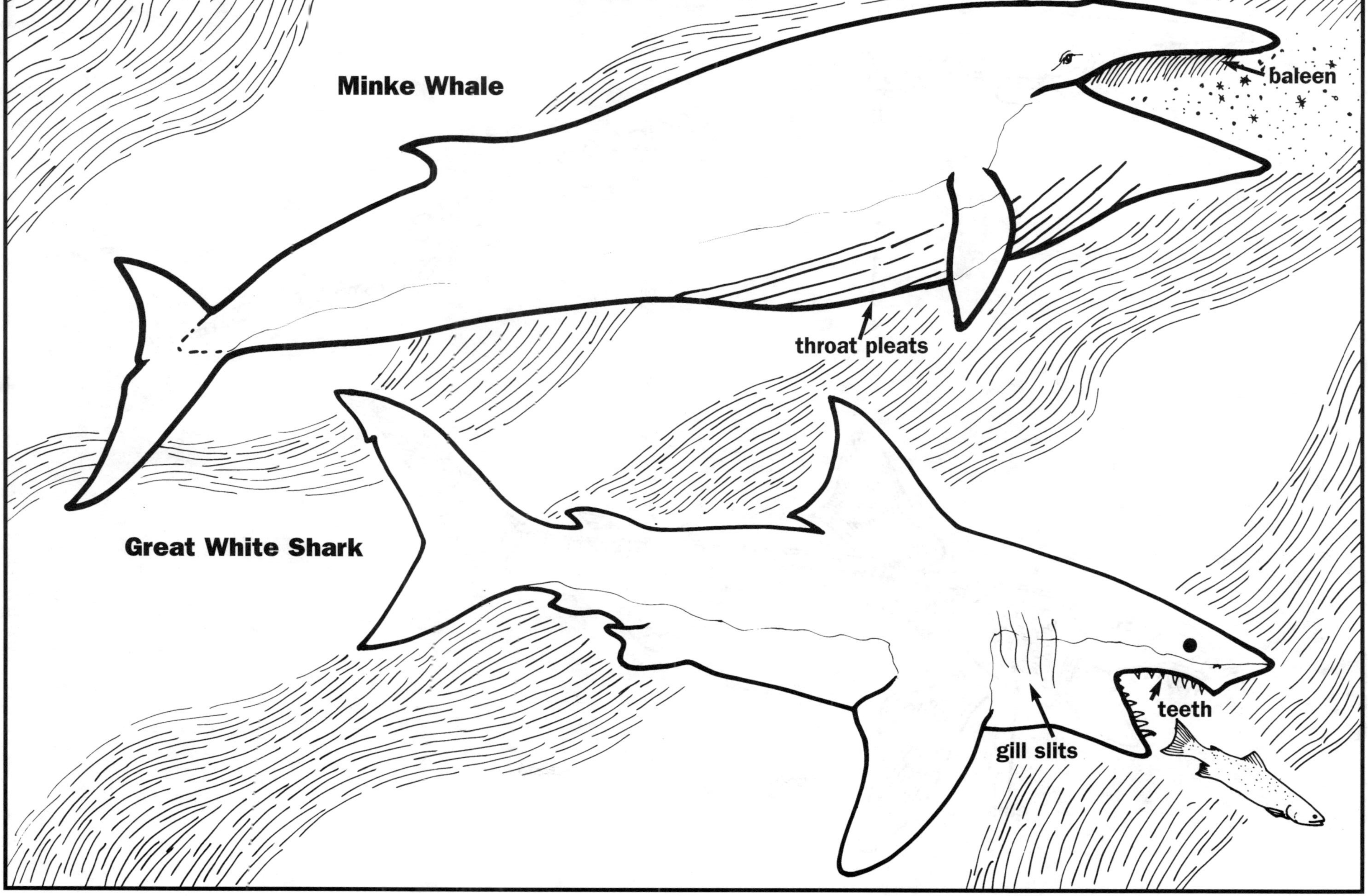

Minke Whale

blowhole

baleen

throat pleats

Great White Shark

teeth

gill slits

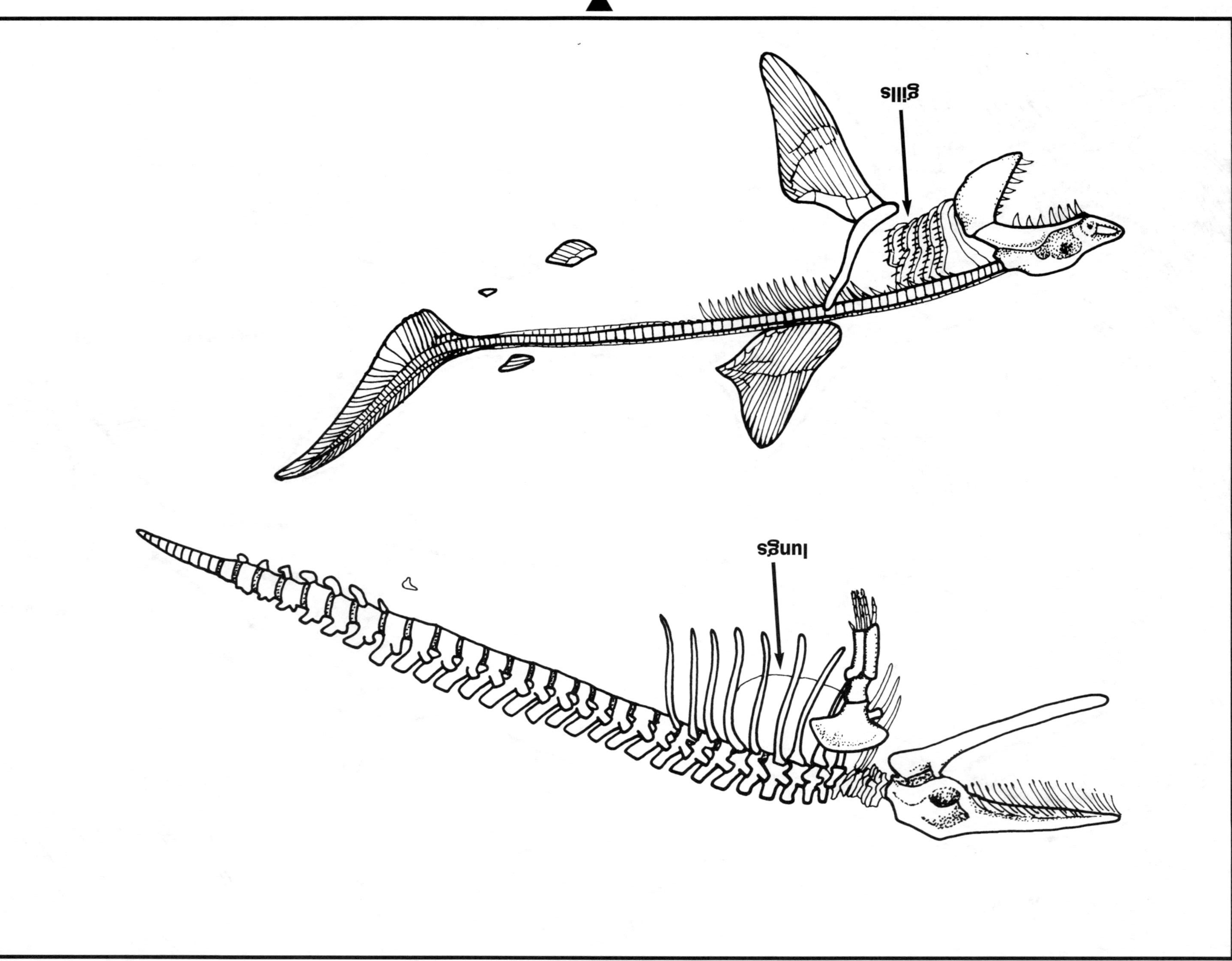
gills
lungs

Name: ____________________

How Are Sharks & Whales Alike?

What to Do

1. Cut out the labels below.

2. Glue them where they go in the circles. (Two have been done for you.)

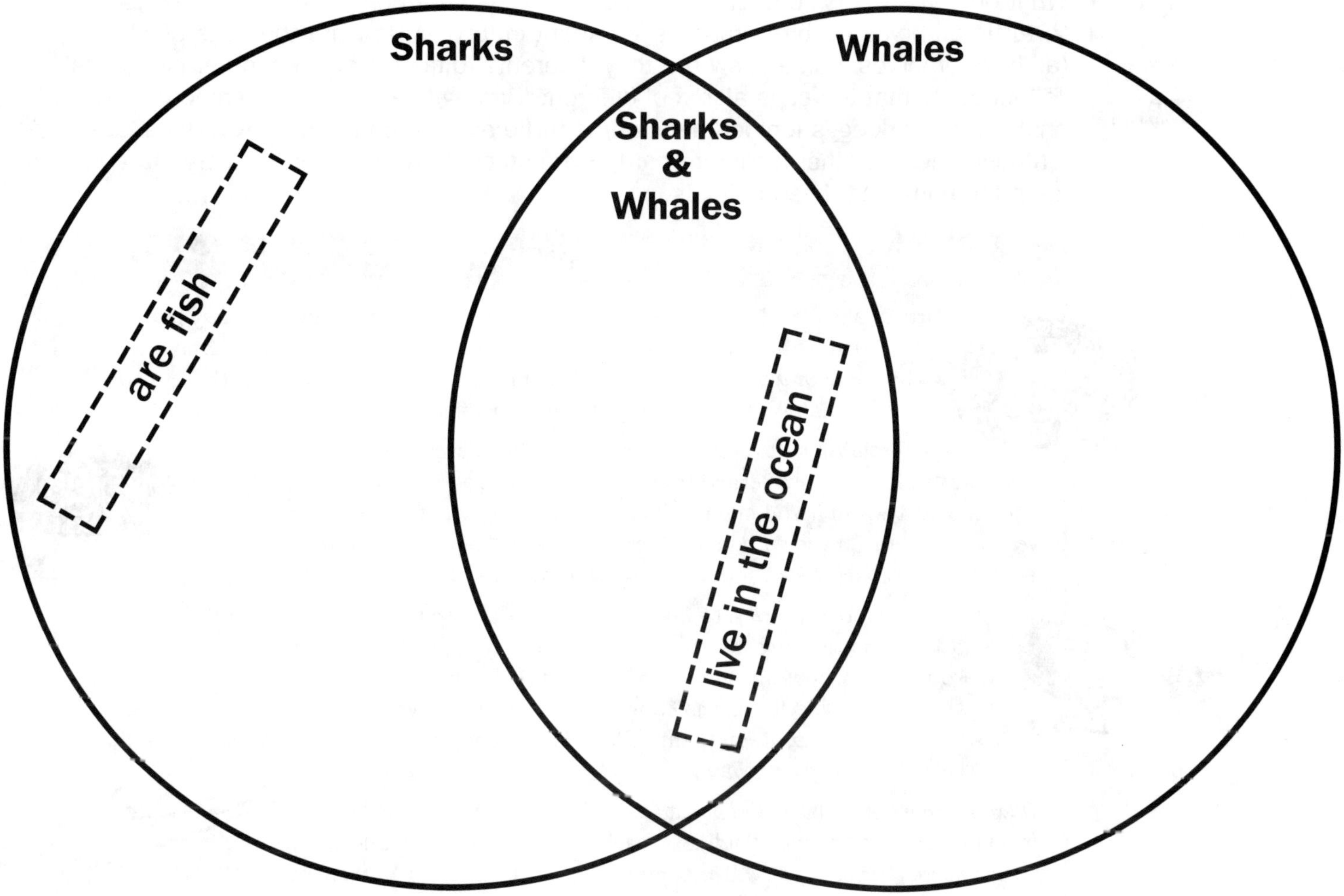

Sharks

Whales

Sharks & Whales

are fish

live in the ocean

LABELS

breathe air with lungs	live in the ocean
are warm-blooded	are mammals
do not care for their babies	have fins
can sense electricity	are fish
nurse their babies	are cold-blooded

Eggs

Using the See-Through (pages 13–14)

Ask students: Where are these four sets of eggs? (1. in a hollow log; 2. buried in dead leaves; 3. in a nest that sits in a tree; 4. in the ocean) What animal baby would you guess is inside each set of eggs?

What do you see inside each set of eggs? (1. snake embryos (unborn animals); 2. turtle embryos; 3. bird embryos; 4. shark embryos) How are the eggs similar? (all have embryos inside) How are they different? (different shapes, different "nests") Tell students that the eggs also feel different. Bird eggs are hard and smooth, whereas snake and shark eggs feel leathery; some turtle eggs, such as the box turtle's, are leathery, whereas other kinds are hard. Invite students to turn over their See-Throughs to learn the names of these animals.

Background

Corn Snake A mother corn snake lays her eggs in a damp, warm, shady place, such as a hollow log, a hole in a tree, or under a pile of leaves. She doesn't tend or guard the eggs. Snake embryos grow for 2 to 3 months. As soon as a baby snake hatches, it glides away to find food.

Box Turtle All turtles bury their eggs on land, even the species that live in water. If they're immersed in water, the embryos will drown. Box turtle eggs are kept warm and damp by the heat given off by the moist, decomposing vegetation in which they are laid. After 1 to 3 months, the eggs hatch. As with snakes, the mother doesn't tend the eggs or care for the babies.

Robin The mother bird sits on her nest to help keep the eggs warm. This incubation shortens the time needed for the embryos to grow, so bird eggs take only 2 to 3 weeks to hatch. A bird embryo has an *egg tooth,* which helps it to peck its way out of the hard shell. Baby robins cannot fend for themselves after hatching—parents must feed and protect them. (Other baby birds, like ducklings and chicks, can find their own food right away.)

Dogfish Shark This shark lays about 20 *eggcases* (the scientific term for shark eggs) at a time. Tendrils attached to the eggcases anchor them to seaweed. Depending on water temperature, it takes 7 to 10 months for the little sharks to develop and hatch. Baby sharks immediately swim away in pursuit of food.

Corn snake eggs are white; the corn snake is orange and red. Box turtle eggs are light pink; the box turtle is light orange and black. Robin eggs are light blue; a robin is brown with an orange head and breast. Dogfish eggcases are brownish-black; the dogfish shark is gray.

Inside an Egg (page 15)

Students use a key to identify the parts of an egg and their functions.

Materials: 1 raw egg and 1 bowl for each cooperative group, and a copy of page 15 for each student

Answers

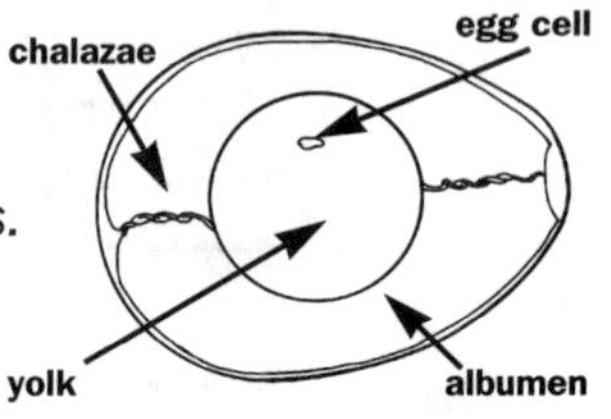

Eggs

What kind of baby animal is inside each set of eggs?

Hold this page up to the light to find out!

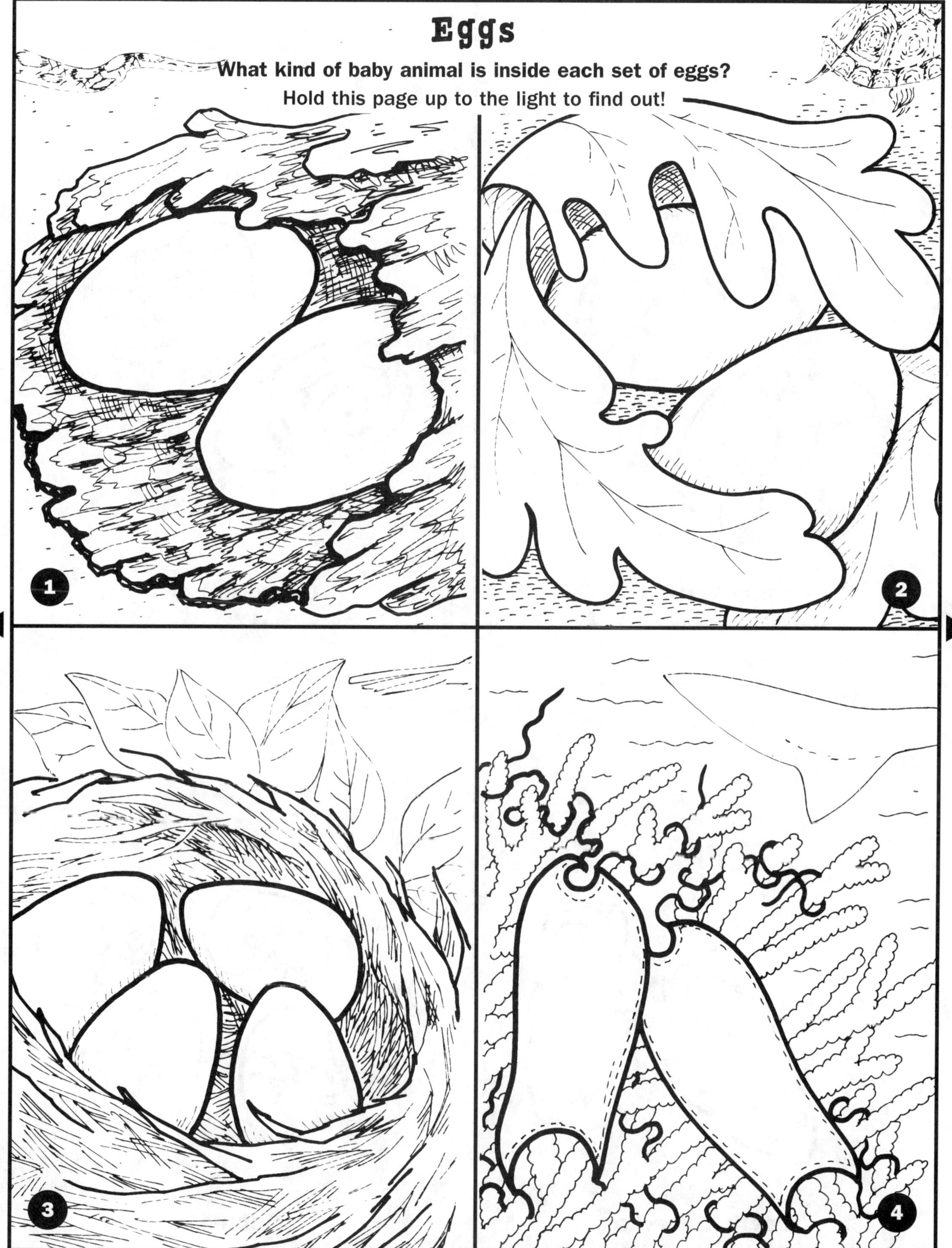

Box Turtles
Corn Snakes
Dogfish Sharks
Robins

Name: ____

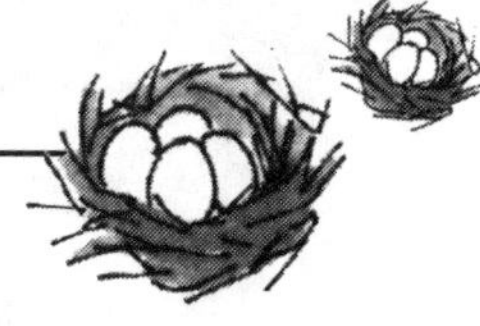

Inside an Egg

How do the parts of an egg help an unborn chick grow? Look inside an egg to find out.

What to Do

1. Crack the eggshell in half. (Ask a grown-up to help.) Carefully pour the inside of the egg into a bowl. Then set the shell aside.
2. Look at the inside of the egg. Draw what you see in the box below.
3. Label the parts. (Use the Key to help you.)

Key

yolk (YOHK) The yellow part that is food for the growing chick.

egg cell The white spot on the yolk. The chick grows from this spot.

chalazae (shah-LAY-zee) The white twisted cords that hold the yolk in place.

albumen (al-BYOO-men) The clear part around the yolk that protects and holds water for the growing chick.

Tyrannosaurus rex

Using the See-Through (pages 17–18)

Ask students: Do dinosaurs like this *Tyrannosaurus rex* live today? (No. Dinosaurs are *extinct*—they have died out.) How do *paleontologists,* the scientists who study fossils, know they once lived? (They found *fossils*, the hardened remains of bones and teeth, and fit them together into skeletons. The skeletons looked like no animals living today, so the paleontologists reasoned that they had discovered animals that lived long ago.)

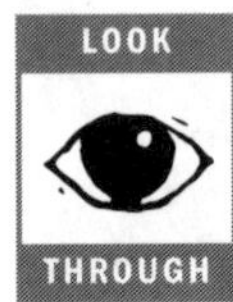

What facts about *Tyrannosaurus rex* might scientists discover by looking at its bones? (Bones show its size and shape; sharp teeth and claws suggest that it ate meat.) What facts about a dinosaur can't be discovered by looking at its bones? (What color it was, exactly what it ate, whether it growled or made other sounds, etc.)

Background

Tyrannosaurus rex lived during the very last part of the Age of Dinosaurs, in the Cretaceous period, from about 70 to 65 million years ago. It lived in what is now Montana, Colorado, Wyoming, North and South Dakota, and Alberta, Canada. At that time, a huge inland sea covered the middle of North America. *Tyrannosaurus rex* would have inhabited leafy forests where lots of ferns and flowering herbs covered the floor and giant redwood trees soared overhead.

There is some debate over whether *Tyrannosaurus rex* was a *predator,* killing its food like a lion, or a *scavenger,* feeding on an already-dead animal such as a buzzard. Some scientists say it was probably both, an *opportunistic feeder.* That means if it found a dead *Hadrosaurus* that died during a long migration, it might eat it. And if it saw a weak, sick dinosaur, *Tyrannosaurus rex* might chase it down.

Whether it was a predator or a scavenger, *Tyrannosaurus rex* needed big, sharp teeth. A large *Tyrannosaurus rex* tooth was 6 inches long. Like other dinosaurs and reptiles today, *Tyrannosaurus rex* grew new teeth every 2 to 3 years, so they were always razor sharp. They did not fall out all at once, but were replaced in a specific pattern at a regular rate throughout its lifetime.

Nobody knows what colors dinosaurs were!
Some scientists think that dinosaurs, like many modern-day birds and a few modern-day reptiles, might have been brightly colored.

Sizing Up Dinosaurs (page 19)

This graph-reading activity lets students compare the sizes of several dinosaurs.

Answers: 1. 30 feet; **2.** *Deinonychus;* **3.** yes; **4.** *Edmontosaurus*

Tyrannosaurus rex

This dinosaur has been extinct for millions of years. How do scientists know about it?

Hold this page up to the light to find out!

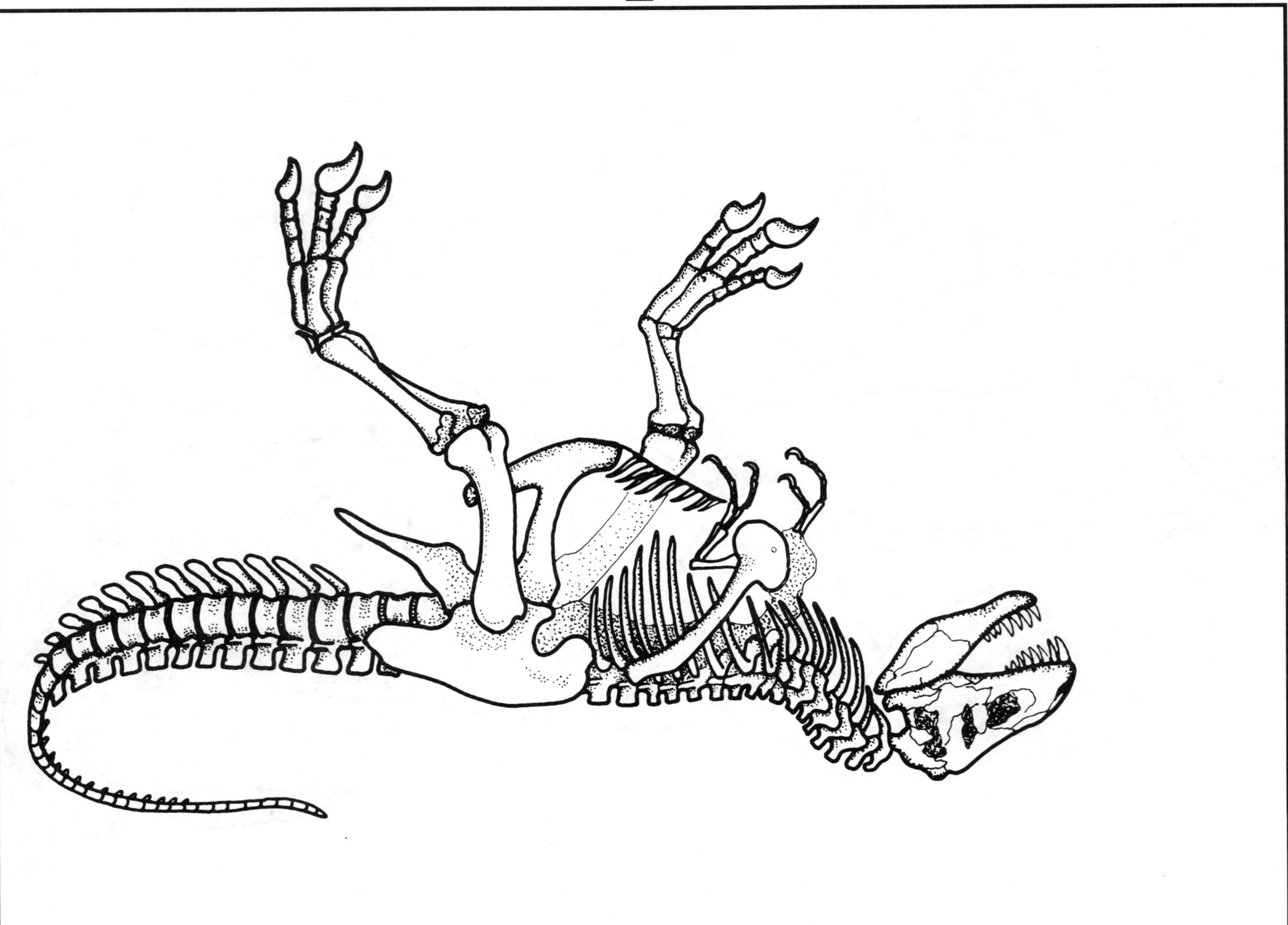

Name: ______________________________

Sizing Up Dinosaurs

**Dinosaurs came in many different sizes.
This graph shows how long some were.
Use the graph to answer the questions.**

KIND OF DINOSAUR

Edmontosaurus
(ed-MON-tuh-SAWR-us)

Triceratops
(try-SER-uh-tops)

Tyrannosaurus rex
(tie-RAN-uh-SAWR-us rex)

Deinonychus
(dye-NON-ik-us)

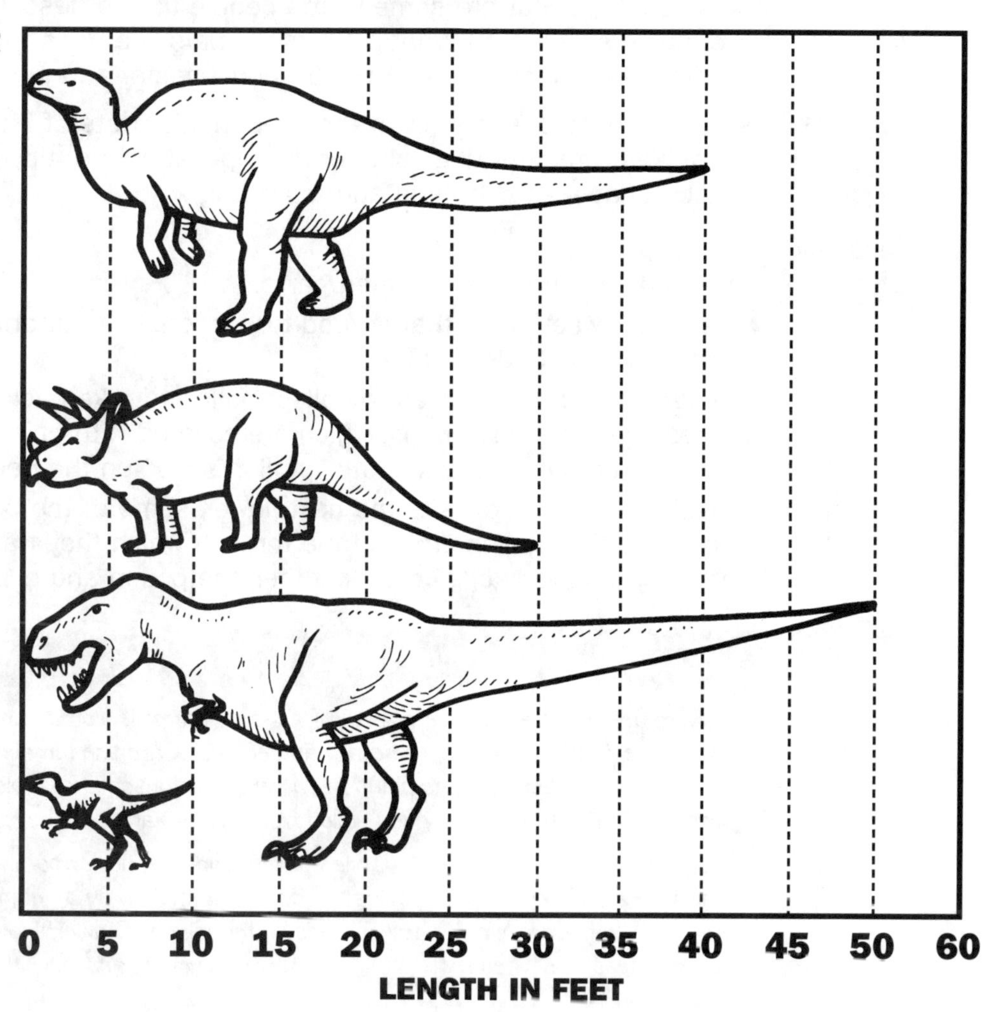

1. How long was *Triceratops*? ______________________

2. Which of these dinosaurs was the shortest? ______________

3. Was *Tyrannosaurus rex* more than 40 feet long? __________

4. Which dinosaur was longer than *Triceratops* but shorter than *Tyrannosaurus rex*? ____________________

EXTRA: Write your own question about the graph. See if a friend can answer it!

____

Tulip

Using the See-Through (pages 21–22)

Ask students: What are some words people use to describe flowers? (pretty, colorful, have a nice smell, etc.) What attracts animals like bees to flowers? (color, fragrance) Why are they attracted to them? (for food—nectar and pollen)

Ask students to describe what they see inside. (lots of little circles—egg cells; long spikes with "hot dog" shapes on top—stamens; tiny dots—pollen grains; tall tube shape—pistil; bees and a ladybug)

Have students find these labeled parts:

◆ **Nectar** A sweet liquid that is food for insects like butterflies and bees and other animals such as birds and bats.

◆ **Pollen** When a bee comes to collect pollen to take back to the hive for food, it brushes against the **stamens**, which are covered with tiny yellow pollen grains. Some of the pollen rubs off and collects in "pollen sacs" on the bee's legs.

◆ **Pistil** When the bee visits the next flower, some of the pollen rubs off on this flower's pistil. The pollen grains grow a tube through which they travel until they reach the **ovary.** There they enter tiny **egg cells.** Together, the pollen and egg cells form seeds.

Background

Flowers are the reproductive organs of a plant. The pollen-producing stamens are the male organ. The pistil, containing the ovary and egg cells, are the female part. Some plants self-pollinate. The pollen from the plant's stamens falls onto its own pistil. However, most flowers cross-pollinate, that is, they use pollen from other plants of their species to grow seeds.

A pollinator is an animal that carries pollen from one flower to another. Each kind of flower attracts a different kind of pollinator. Bats and moths fly at night. They look for nectar in bright white flowers, which they can see by moonlight. Butterflies like red, orange, and yellow flowers. Bees don't see red, so they go for blue, yellow, and purple flowers.

Tulips are red, light or bright pink, yellow, or purple. Their leaves are green. The ladybug is red with black dots. The bees' bodies have yellow and black stripes. Yellow pollen covers the large bee's back legs. The sulphur butterfly is yellow or orange.

Inside My Flower (page 23)

Students identify the parts of a flower.

Materials: simple flowers such as tulips, apple blossoms, or lilies (florists often provide slightly aged flowers at no charge); Tulip See-Through (pages 21–22); a copy of page 23 for each student; clear tape

Pass out flowers and the reproducible and guide students through the activity. (Hint: To pick up pollen with tape, brush the tops of the stamens, called *anthers,* with the sticky side of the tape.) Wrap up by asking students to look at the back of the Tulip See-Through. How is the tulip shown similar inside to the flowers they took apart? How is it different?

Flower

What's inside this tulip?

Hold this page up to the light to find out!

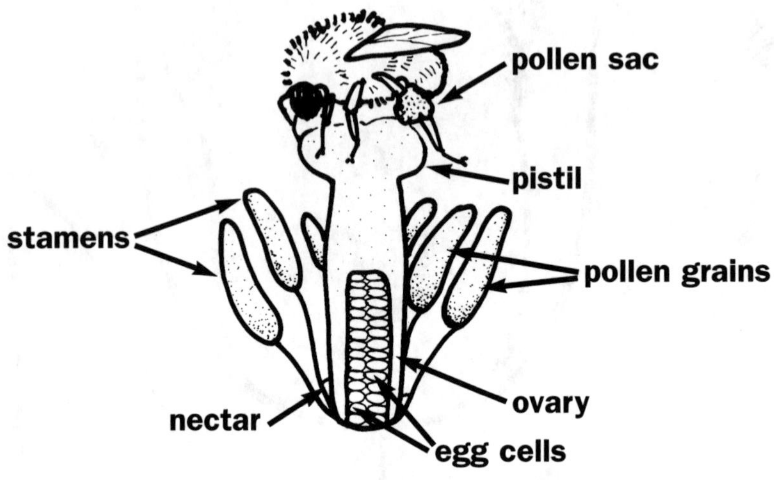
pollen sac
pistil
stamens
pollen grains
ovary
nectar
egg cells

Name: ____________________

Inside My Flower

Record what you see here.

1. PETALS How many petals does your flower have? __________

What color are the petals? __________

◀ Draw one of the petals here.

2. STAMENS

▼ Draw the stamens here.

Use tape to pick up some pollen.

▼ Tape the pollen here.

3. PISTIL Touch the top of the pistil.

Is it sticky? __________

What usually sticks to the top of the pistil?

____

◀ Draw the pistil here.

Hidden Homes

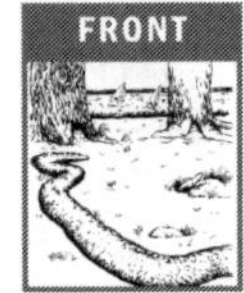

Using the See-Through (pages 25–26)

Ask students: What signs of animal life do you see? (a fluffy tail that belongs to some kind of animal, a tiny nose and tail poking out of tree hollows, a tunnel of pushed-up dirt) Where do you think these animals live? (in tree hollows and underground)

What hidden animal homes do you see? (mice in their nests, mole and worms in their tunnels, a chipmunk in its burrow)

Background

◆ If you see a zigzag line of fresh dirt on a lawn, it's probably the work of a mole. It pushes the dirt straight up with its big front claws, then checks back later to see if any insects or worms fell down into its tunnel. Because it lives in the dark, a mole has no need to see—it has very poor eyesight. Sensitive whiskers help it find its way.

◆ A worm is one of nature's great recyclers. It eats dead plant material and leaves behind rich soil as a waste product. As worms push through soil, they also help aerate it. This is helpful to growing plants.

◆ Mice sometimes make their nests in the hollows of trees or in underground burrows. They may even recycle empty bird nests! They hide nuts, seeds, and grains in their burrows or in hollow trees. They retrieve their food, not by remembering where they buried it, but by sniffing it out with their sensitive noses.

◆ A chipmunk stuffs seeds and nuts in its expandable cheek pouches, then stores its bounty in its underground burrow.

Mouse: orange-brown; mole: black or brown; worm: pinkish-brown; robin: brown with an orange head and breast; chipmunk: tannish-black with a black stripe on its back

Creepy Crawler Count (page 27)

Students observe tiny animals that live in soil.

Materials: a lamp with a flexible neck, a strainer with large holes, a large bowl lined with a wet paper towel, soil with decomposing plant matter from outside (about two cups), hand lenses, and a copy of page 27 for each student

1. Set up the materials as shown. Put the soil in the strainer. Position a lamp about 6 inches above the strainer and leave the lamp on overnight.

2. The next day, lift out the strainer. Use hand lenses to look in the bowl and under the towel for "creepy crawlers"—insects and other arthropods, earthworms, etc. (Safety note: If there are dangerous bugs in your area, check the bowl before allowing children to handle it.) If no animals appear, leave the lamp on for two or three nights. (Keep the paper towel damp.)

3. Have students record the creepy crawlers they see on their charts. Then return the bugs to their natural habitat.

DISCUSS: What did the heat from the lamp do to the soil? (started to dry it out) Why did the creepy crawlers fall through the holes in the strainer? (They were looking for moister soil.) Why did we use a wet paper towel? (to provide a comfortable place for the bugs)

Hidden Homes

What animals live in this park?

Hold this page up to the light to find out!

mouse

mouse

chipmunk

mole

grub

beetle

worm

Name: ____________________

Creepy Crawler Count

What animals did you find in your soil?
Make a ✔ for each creepy crawler you see.

Mite	Insect Larva	Springtail	Millipede	Earthworm	Pill Bug

(Animals shown are not their real size.)

Draw other creepy crawlers you see below.

Cave

Using the See-Through (pages 29–30)

Ask students: What evidence do you have that there's a cave under this field? (dog peering into hole in the ground, bats flying from behind a hill)

Who lives in this cave? (bats, fish, salamanders) What else do you see in the cave? (shafts of rock on the floor of the cave and hanging from the ceiling, rock columns joining floor to ceiling, and an underground stream)

Invite students to think about conditions in the cave.

Ask: Where do you think the water comes from? (from rainwater seeping in from above, underground rivers) Is there light? (only near the entrance) Are there plants? (No, green plants need light to grow.)

◆ Of the animals that live in this cave, only one kind ever leaves. Which kind is it? (bats) Why do you think bats leave? (Guide students to understand that bats sleep in the cave by day and go out at night in search of food. Big brown bats, for example, hunt for flying insects. How do they find find their way and nab bugs in the dark? Bats use *echolocation*—they send out high-pitched squeaks and listen for the echoes to return. The quicker the echo, the closer an object is.

◆ Other cave animals never leave the cave. They feed on the remains of dead cave animals, food that is carried into the cave by other animals or by water, and even bat droppings, called *guano.*

Background

Cave rock is called *dripstone,* because it forms when water drips down from the cave's ceiling. As the drips evaporate, small rings of dissolved rock are left behind. These rings "grow" into a *soda straw* formation with a hollow center. This is a very slow process—it takes a hundred years of dripping to make only one inch of a soda straw! Over time, soda straws grow into huge hanging *stalactites.* When water drops to the cave floor, *stalagmites* grow upward. A *column* forms when a *stalactite* and a *stalagmite* grow into each other.

In a cave's deep, dark interior live colorless creatures that have no eyes. In the dark, there is no need to see; and colors, which serve to defend, attract, or camouflage, have no useful purpose here. Instead, cave animals have a sharp sense of smell and touch.

Cave walls are yellowish-brown.
Bats are brown, gray, or black.
The other animals are clear or pinkish-white.

EXTENDING LEARNING

Try on a Bat's Ear! (page 31)

Students make a model of a bat's ear to investigate how large outer ears help bats hear well.

Materials: scissors, tape, a radio or audiotape with a relatively constant volume for students to listen to, and a copy of page 31 for each student

Cave

What's under this field?

Hold this page up to the light to find out!

stalactite

bat

column

cave fish

stalagmite

cave
salamander

Name: ____________________

Try on a Bat's Ear!

Why do bat's have such big ears?
Try this and hear for yourself!

What to Do

1. Cut out the Bat Ear pattern along the thick solid lines.

2. Curve the paper as shown. Line up the small bats.

3. Ask a friend to put tape along the edge.

4. Follow the steps on the back of the ear.

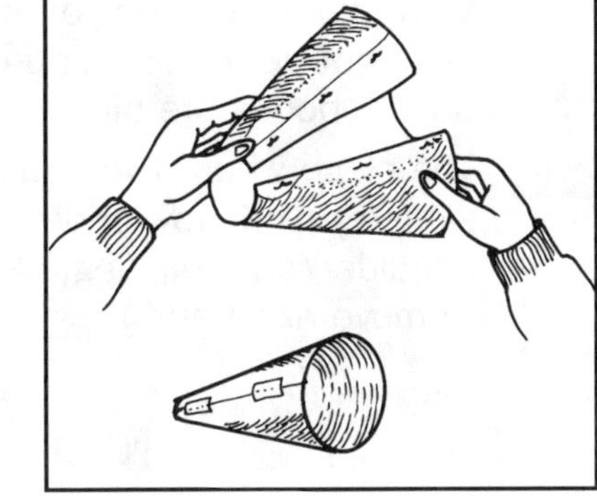

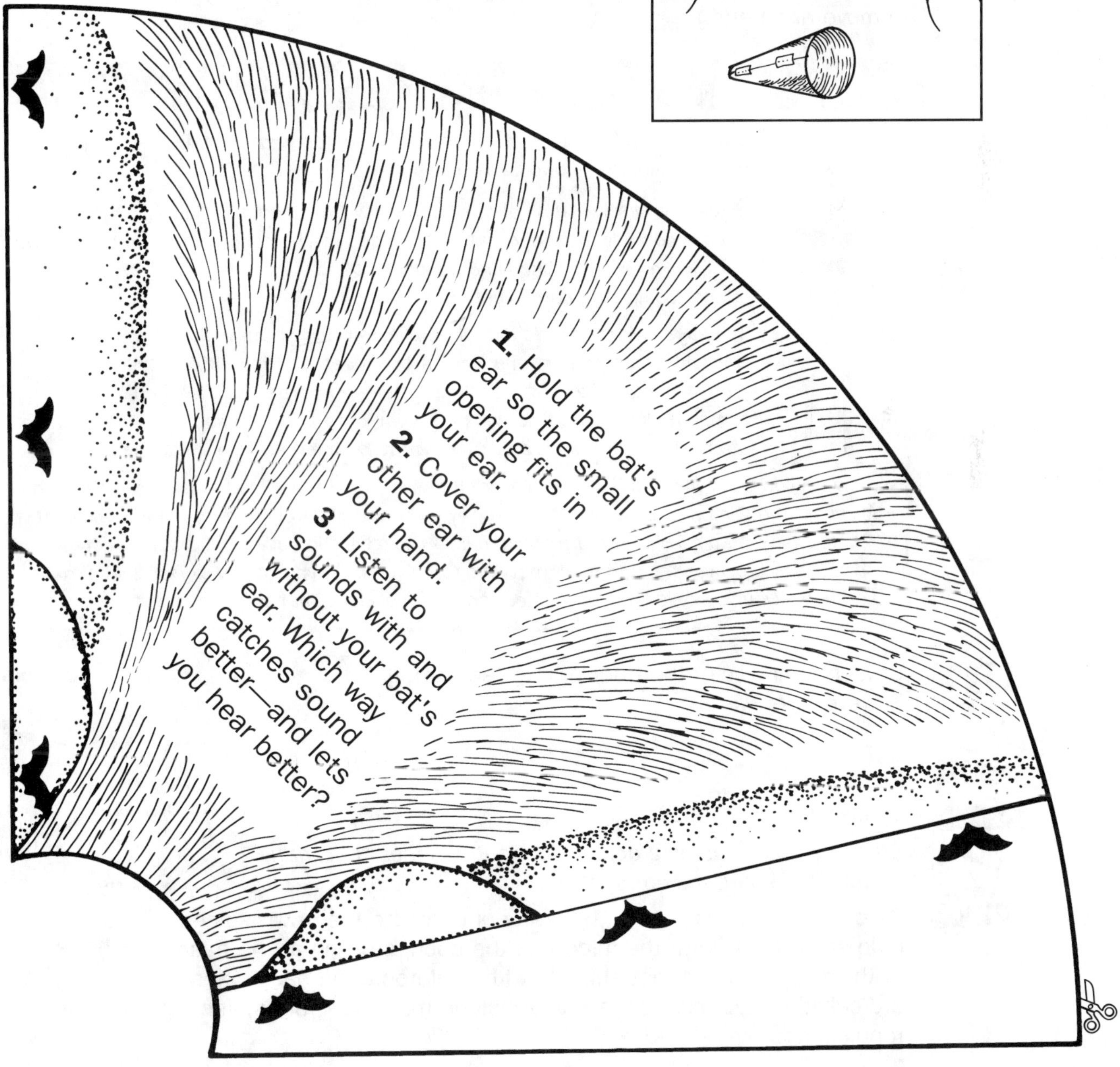

1. Hold the bat's ear so the small opening fits in your ear.

2. Cover your other ear with your hand.

3. Listen to sounds with and without your bat's ear. Which way catches sound better—and lets you hear better?

Human Skeleton

Using the See-Through (pages 33–34)

Ask students: What if you didn't have a skeleton—would you be able to stand up? (No, you'd be a soft, lumpy bag of skin, with organs and muscles inside.)

What differences do you see between the skeletons of the adult and the child? (size, shape of pelvis, skull, rib cage) Do bones grow? (yes) How do you know? (because the adult's bones are bigger)

The place where two bones meet is called a *joint*. Ask: Where do you see joints on the skeletons? (Elbow and knee joints are easy ones to identify. Others include ankle, wrist, shoulder/arm, hip/leg, skull/neck.) What do joints do for you? (They allow the skeleton to move and bend.)

Background

A skeleton has three main functions: it holds our bodies up, protects delicate organs, and, with the help of muscles, allows us to move.

A baby is born with all of the bones it will have as an adult (206) although all are not fully developed at birth. Most of the bones in our skeleton stop growing by the time we are 16 to 20 years old, but some skull bones do not completely harden until age 30.

A baby's skeleton isn't just a smaller version of an adult's—it has very different proportions. One of the most striking differences is between skull size and overall height. A newborn's body is about 4½ times as long as its head. A five-year-old's body is about six times as long as its head. An adult's body is about 8 times as long as its head.

Living cells in the bone secrete minerals which make them hard. Blood vessels run through the bones, bringing nutrients and oxygen to living bone cells and removing waste. The center of bones have a spongelike texture which makes them light in weight. At a bone's core is marrow, which stores fat and manufactures red blood cells that carry oxygen and carbon dioxide, white blood cells that fight infection, and platelets that help blood clot.

Students may color the front of the See-Through however they wish.

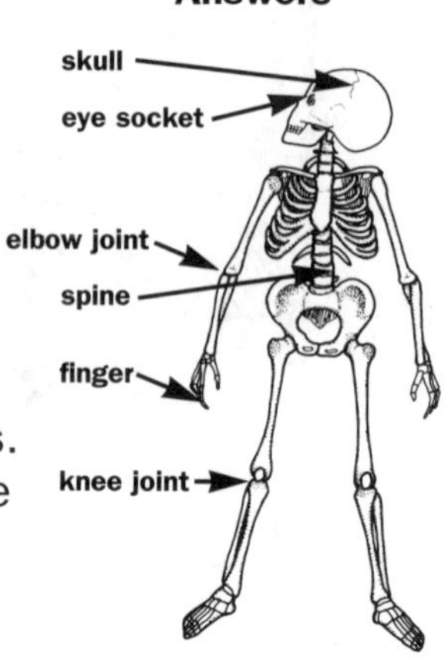

EXTENDING LEARNING

Connect the Bones (page 35)

Students do a matching activity to identify bones that many vertebrates share in common.

Before starting this activity, help students locate their own spines. (The spine runs down the middle of the back, starting at the base of the skull.) Explain that animals with skeletons and backbones are called *vertebrates*. Vertebrates include mammals, birds, fish, reptiles, and dinosaurs.

Human Skeleton

How does your body look inside?
Hold this page up to the light to find out!

TYRANNOSAURUS
REX

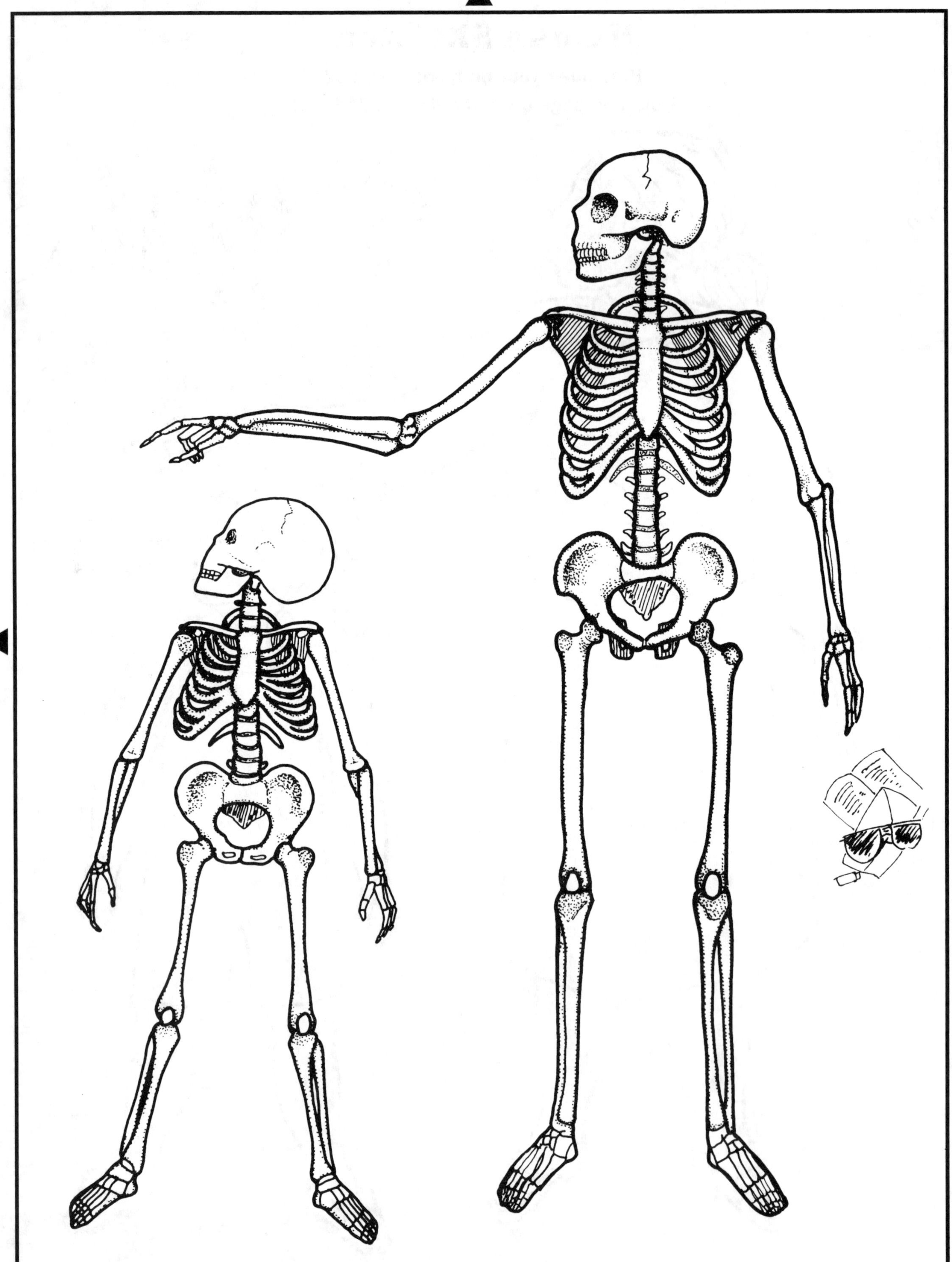

Name: ______________________________

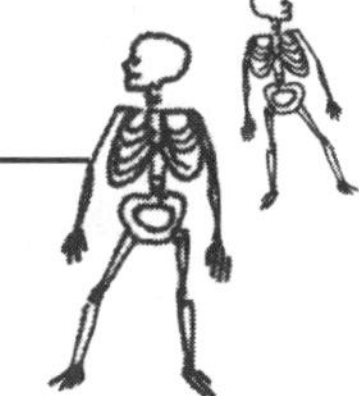

Connect the Bones

What do you and a frog have in common?
A lot of the same bones!

Look at each bone label. Where does it point to on the frog skeleton? Where should it point to on the human skeleton? Draw a line to show your answer.

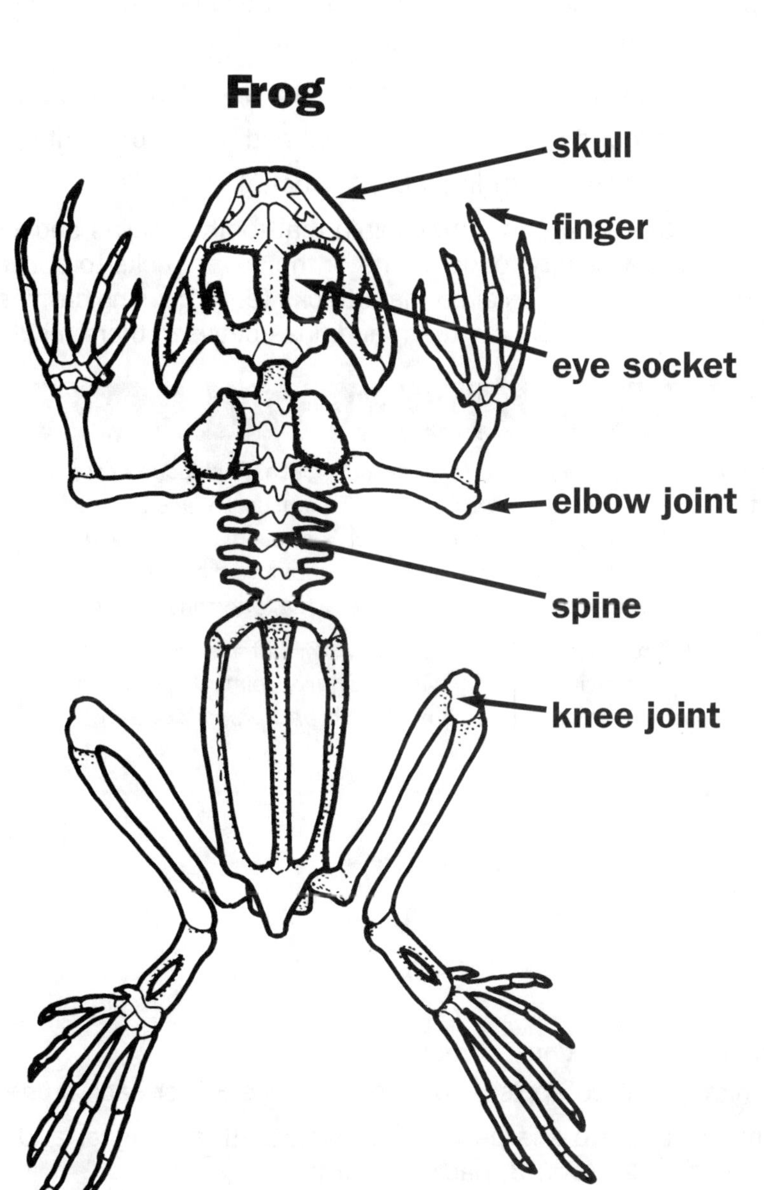

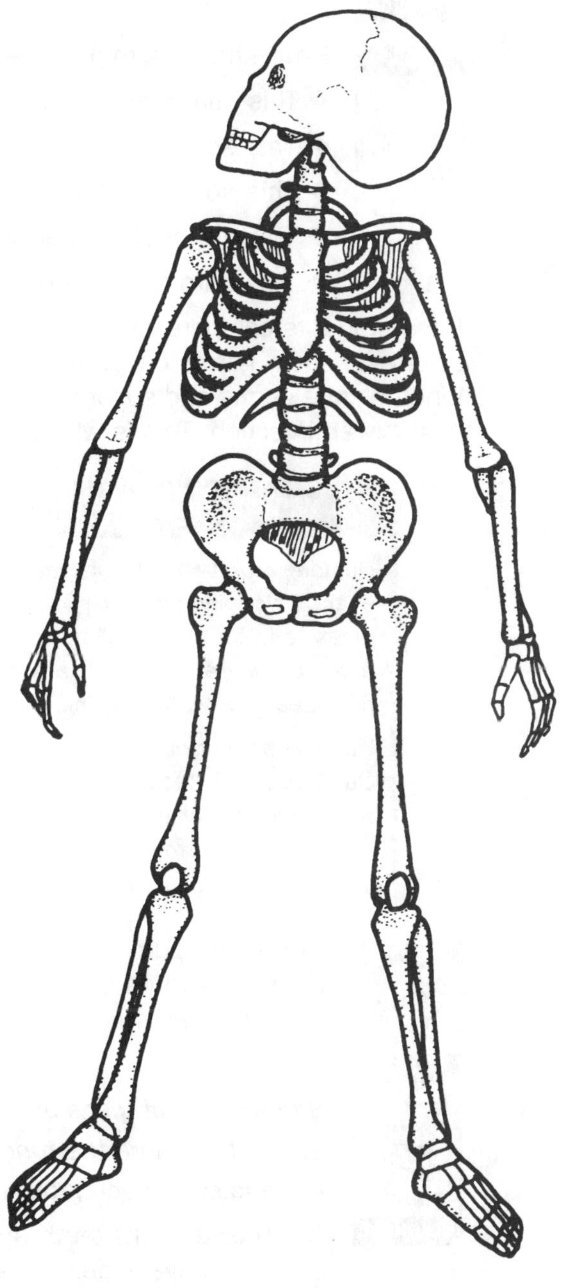

Healthy & Decayed Tooth

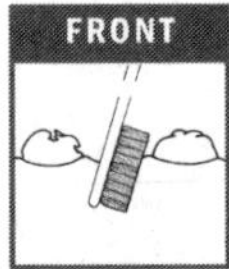

Using the See-Through (pages 37–38)

Ask: What differences do you notice between these two teeth? (The *crown* of the healthy tooth is solid; the decayed tooth has cracks and holes, called *cavities.*) What do you think is inside each tooth? (Encourage students to explain the reasons for their ideas.)

Are most of these teeth above or below the gums? (below—The roots extend down into the jawbone.) What differences do you notice between the two teeth now? (The decayed tooth has a dark line, indicating infection, running down from the cavity to the base of the root.)

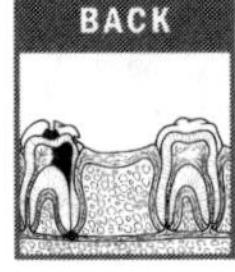

Have students match these descriptions with the labeled parts on the healthy tooth.

- ◆ This hard outside covering protects the tooth from damage and keeps out germs. (enamel)
- ◆ This soft, pink part protects the teeth, its roots, and the jawbone. (gums)
- ◆ This soft part, found in the middle of the tooth, is full of nerves and veins. (pulp)
- ◆ This material, harder than bone, is found between the enamel and the pulp. (dentin)
- ◆ This hard material holds the roots of the tooth in place. (jawbone)

Have students compare the healthy tooth with the decayed tooth. Explain that tooth decay is caused by colorless, sticky germs called *plaque.* When we eat sugary foods, some of the sugar sticks to our teeth. When plaque germs and sugar mix, they turn to acid—a substance that attacks teeth, causing cavities and possibly an infection. This is why it is so important to brush our teeth and floss between them regularly.

Background

Humans grow two sets of teeth: *deciduous* (or baby teeth) and *permanent* teeth. Deciduous teeth start developing before birth but don't make an appearance until around 6 months. By age 2½, most toddlers have a full set of 20 baby teeth. Eventually, the roots of these teeth dissolve, and at around 6 years old, a child's baby teeth begin to fall out. Over time, permanent teeth move in to fill those gap-toothed grins. By about 21 years, most young adults have all 32 permanent teeth.

Humans are *heterodonts,* that is, we have different kinds of teeth adapted for different purposes. Our *incisors* (located in front) have a straight cutting edge for biting. Sharp, pointed *canines* are also designed for biting and tearing off food. In the back of our mouths, broad *premolars* and *molars* are built for crushing and grinding food.

Gums and pulp are pink.
Enamel is white or yellowish.
Dentin is yellowish tan

Bite and Chew! (page 39)

Students explore the functions of different kinds of teeth.

Materials: for each group—a mirror, a sliced apple, and a copy of page 39 for each student

Answers: **1.** front teeth are smaller and are sharper on top, back teeth are bigger and have wider surfaces on top; **2.** front; **3.** back; **4.** hard;
Think About It: front—biting; back—chewing

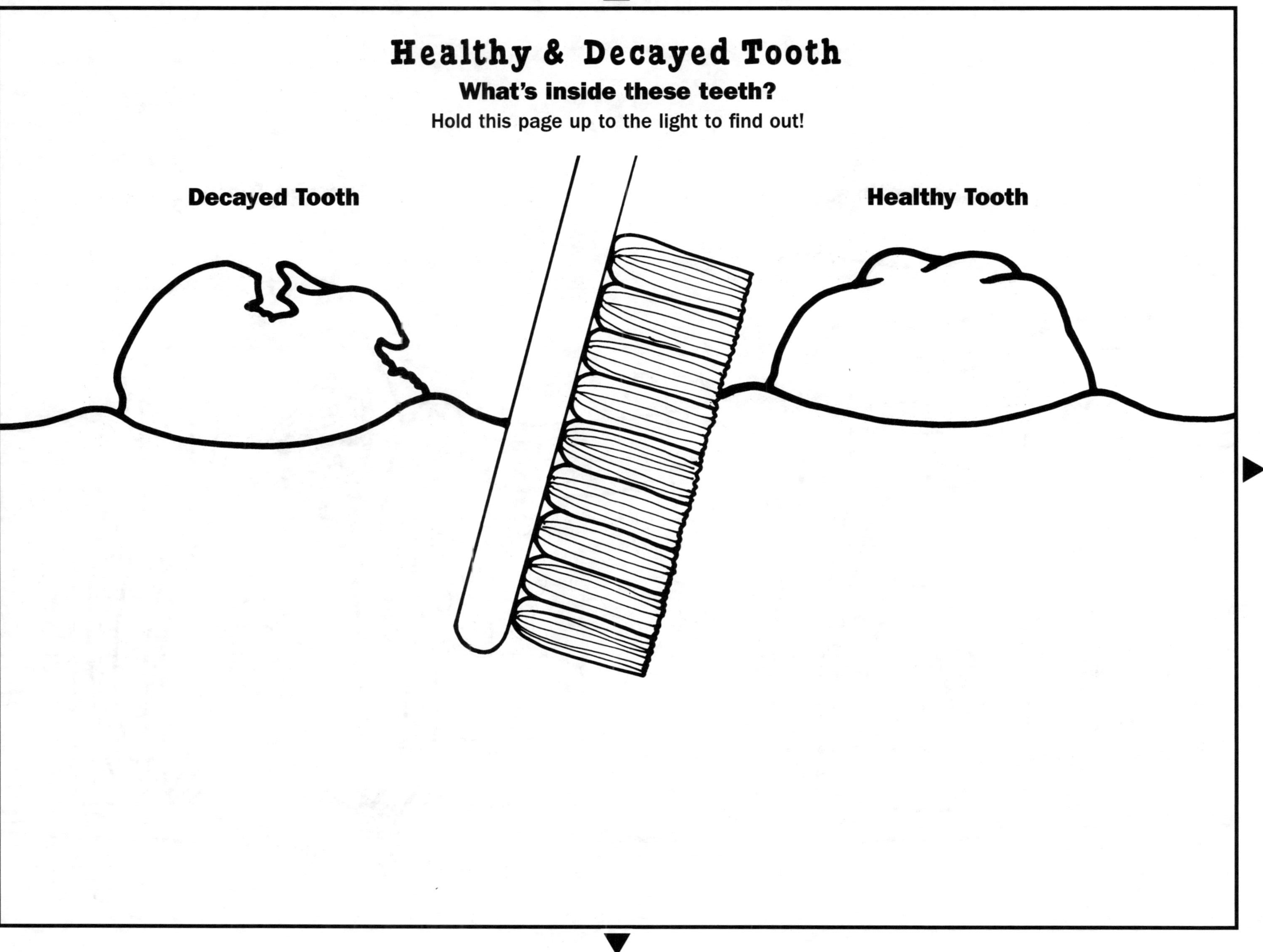
Healthy & Decayed Tooth
What's inside these teeth?
Hold this page up to the light to find out!
Decayed Tooth
Healthy Tooth

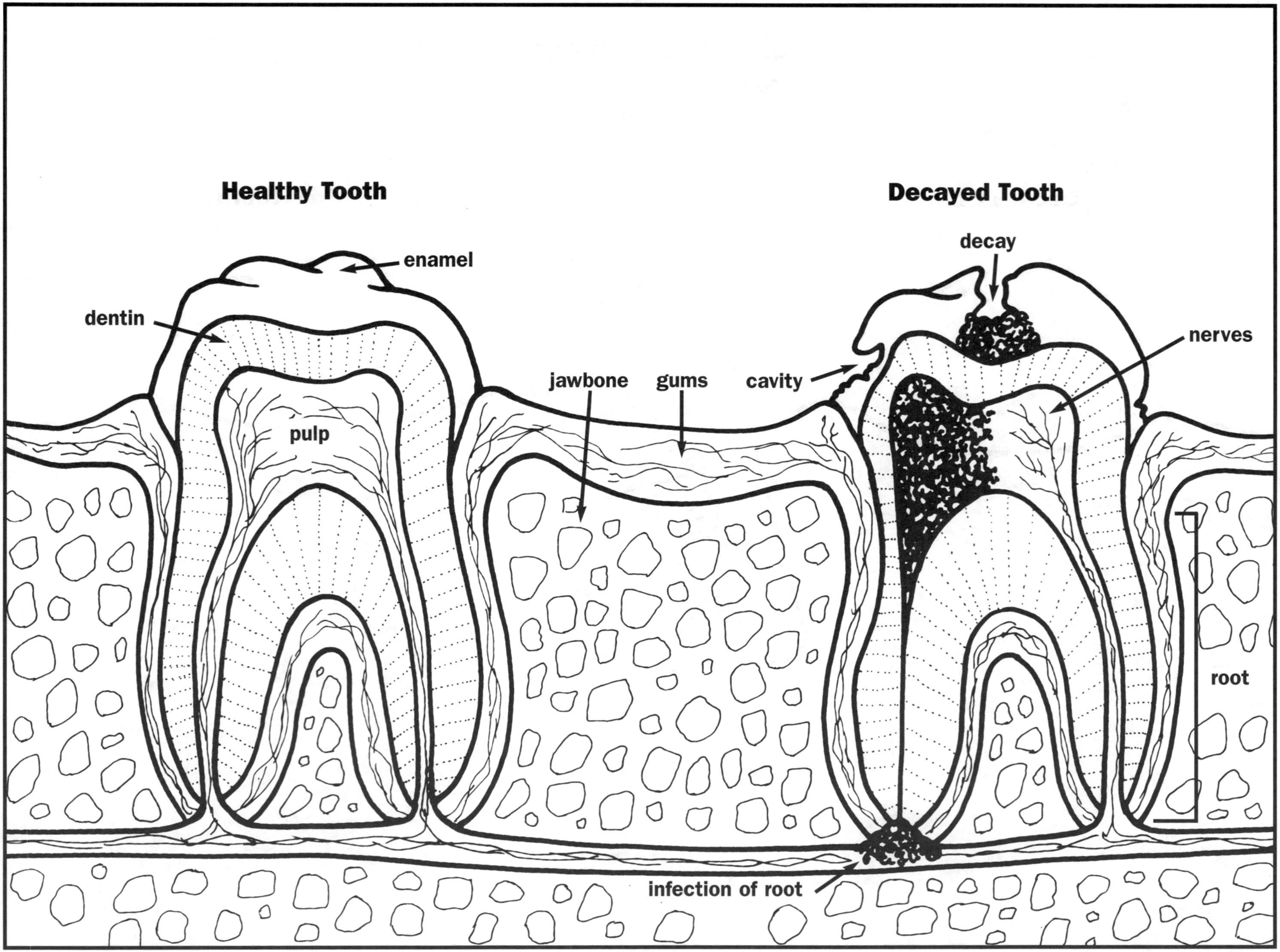

Healthy Tooth

enamel
dentin
pulp

Decayed Tooth

decay
nerves
cavity
gums
jawbone
root
infection of root

Name: ____________________

Bite and Chew!

Your teeth tear and grind food into smaller pieces. What jobs do different teeth have? Find out here.

What You Need

- ◆ mirror
- ◆ apple slice

What to Do

1. Look at your teeth in a mirror.
How are your front teeth different from your back teeth?
Write your answer here: ____________________

2. Take a bite of your apple slice.
Which teeth did you use—the front or back? ____________________

3. Chew the apple.
Which teeth are you using now? ____________________

4. Take a second bite of your apple slice.
This time, chew it with your front teeth only.
Is it easy or hard to chew this way? ____________________

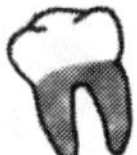

Think About It: Your front and back teeth have different jobs.
What are they? ____________________

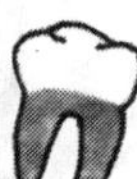

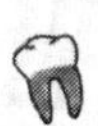

Volcano

Using the See-Through (pages 41–42)

Challenge students to find these volcanic products on the drawing.

◆ In a volcanic eruption, *ash* can be blown more than three times as high as an airplane can fly. (clouds of ash in the sky over the volcano's opening)

◆ Melted rock, called *lava,* can flow like a river. Lava can be runny and move fast, or thick and slow-moving, like honey. (lava running down the sides of the volcano)

◆ Huge blobs of lava can burst out of a volcano. As they fly through the air, they cool and harden into rocks called *volcanic bombs.* (chunks of lava in the air around the volcano)

Where do the ash, lava, and volcanic bombs come from? (from underground, beneath the volcano) Ask students to point to the shaft at the bottom of the picture. Explain that when lava is still underground, it is called *magma.* Magma forms deep underground, where great pressure and hot temperatures melt rock.

Before a volcano erupts, magma rises up and collects in the volcano's *magma chamber.* (Tell students to move their fingers up to the magma chamber.) There, the melted rock is put under pressure from other magma pushing up from below.

What happens when too much pressure builds up? Have students compare this to shaking a can of soda, then opening it up. What happens? As the pressure is released, the gases and liquid in the can burst out in the same way that lava and ash escape through a crack or weak spot in the earth's crust, creating a volcano.

Tell students that the light and dark areas inside the volcano's cone are layers of hardened lava and ash that settled after previous explosions.

Background

Volcanoes erupt, ooze lava, and drop ash over and over again. Eventually, the layers of ash and lava form what we recognize as the common cone shape of a volcano. Most of the gas that escapes from volcanoes is steam (water) and carbon dioxide. Earth's oceans and atmosphere were created by gases that escaped from volcanoes 4.5 billion years ago as the planet formed.

Volcanoes can be quiet, or dormant, for hundreds of years. Though no one can predict when a volcano will erupt, some volcanoes have fairly regular eruption patterns. Mount St. Helens, in the state of Washington, erupts about every 150 years. In Hawaii, active volcanoes Kilauea and Mauna Loa erupt every 2 to 3 years.

Lava is bright reddish-orange as it spews out of a volcano. The outer part of the flow darkens as it hits cool air. Ash is gray. Volcanic bombs are dark gray.

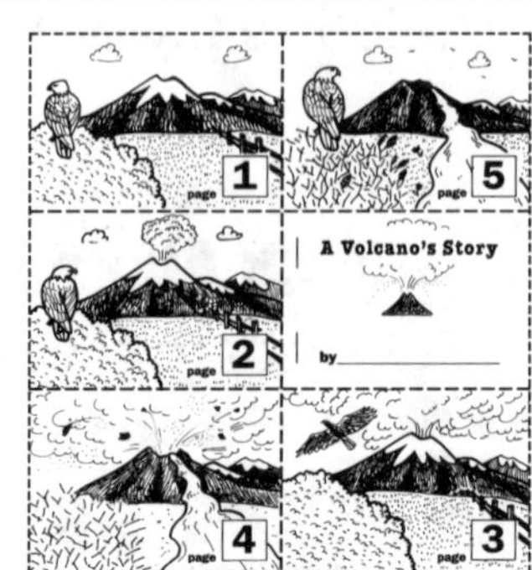

Volcano Mini-Book (page 43)

Students sequence the stages of an erupting volcano.

Materials: scissors, pencils, crayons or markers, and a copy of page 43 for each student

Volcano

What's inside a volcano?
Hold this page up to the light to find out!

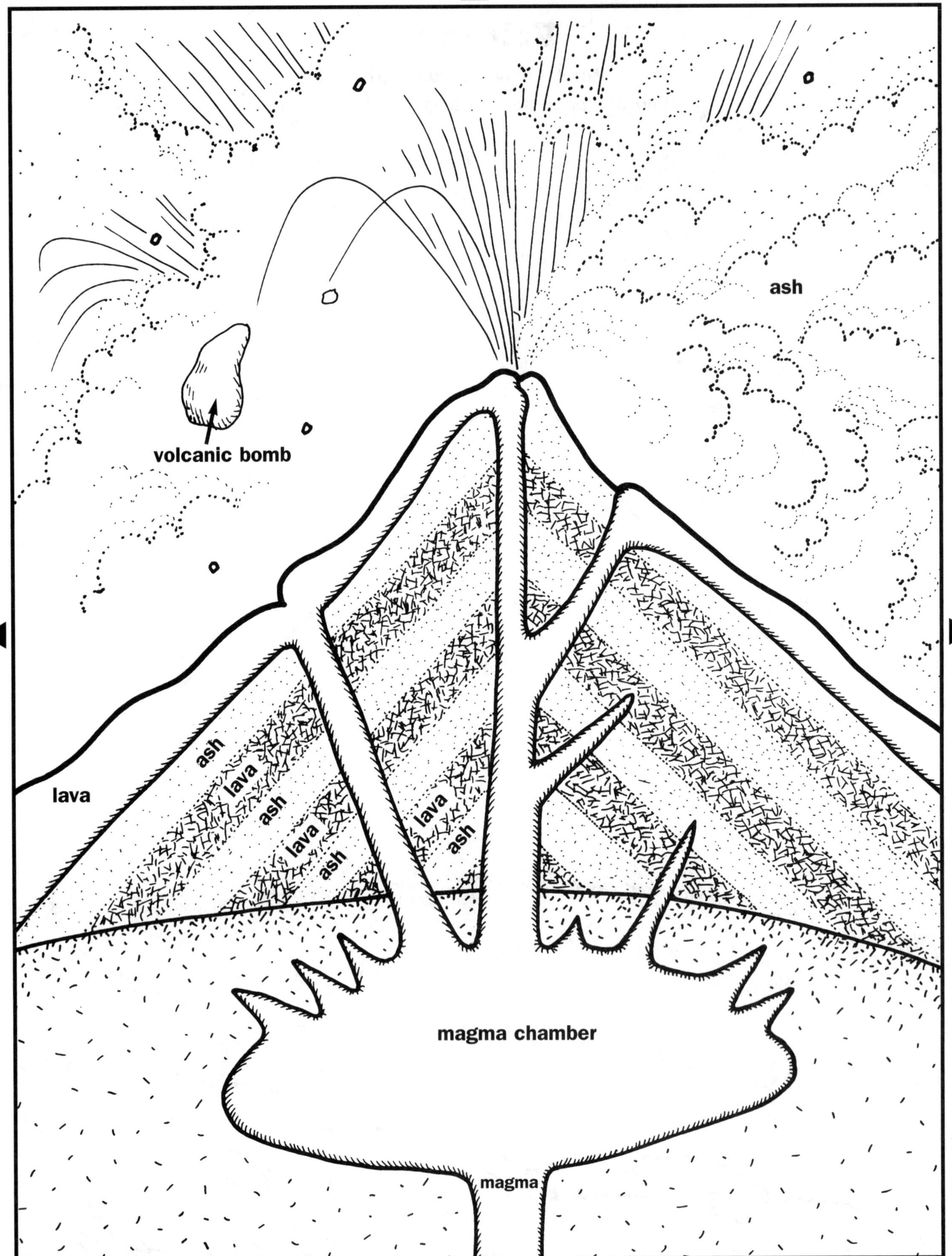
ash
volcanic bomb
lava
ash
lava
ash
lava
ash
lava
ash
magma chamber
magma

Name: ________________________________

A Volcano's Story Mini-Book

1. The pages of this book show a volcano as it erupts. But the pages are all mixed up. Number the pages from 1 to 5.

2. Cut the pages apart and put them in the right order. The page that says *A Volcano's Story* goes on top. Staple the pages on the left side to make the book.

3. On the blank pages in your book, write a sentence to go with each picture.

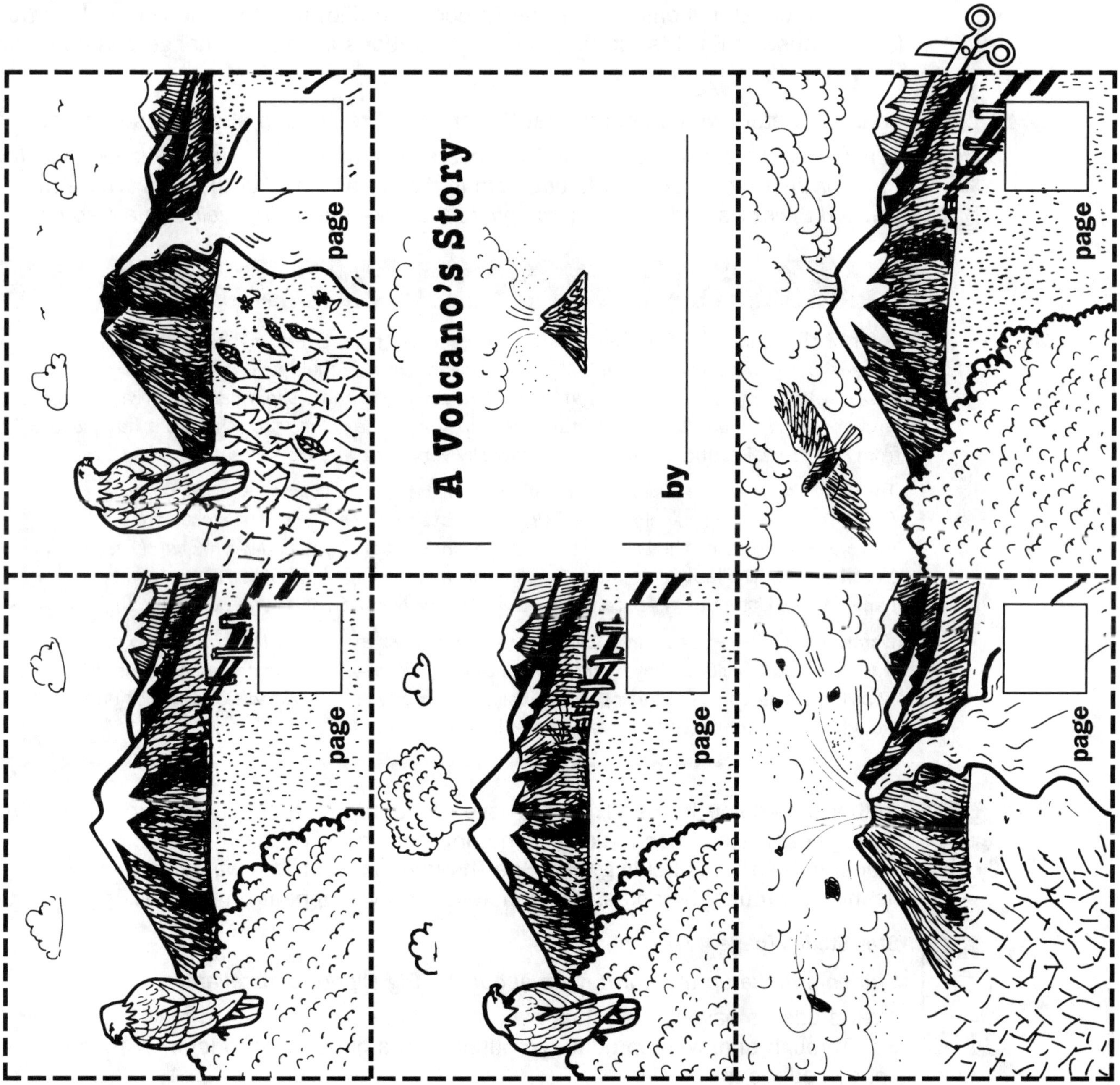

page

A Volcano's Story

by ______________

page

page

page

page

Constellations

Using the See-Through (pages 45–46)

Tell students: People have looked at stars for thousands of years. To explain the patterns and movements of the stars in the sky, people imagined that groups of stars were pictures, and they made up names for them and stories about them. These star pictures are called *constellations*.

One group of people, the ancient Greeks, imagined that these two star groups were two bears. Ask students if they can guess where the bears are—one big and one little. (Encourage students to try to identify heads, tails, legs, etc.)

Note: The constellations on the See-Through are shown in their positions during the fall. Because of Earth's rotation, the constellations appear to change position during the year.

Share this story with students that the ancient Greeks told about these bears.

A mother and her child were turned into bears by a magic spell. The Greek god Zeus felt sorry for them, so he decided to put them in the sky where all of the gods lived. Zeus grabbed their tails and tossed them up into the sky—that's why their tails are so long.

Background

Most of the names for the constellations we see are from Greek myths first told 2,000 years ago. By watching the night sky, the ancient Greeks could tell what time of night it was, or what time of year. Star patterns were their compass, too, when they traveled by land or sea. Although the stars in a constellation appear close together as a flat picture from our viewpoint on Earth, they are actually very far away from one another in space.

The Great Bear includes one of our most recognizable star patterns—the Big Dipper. (The bear's tail is the handle of the Dipper.) Two stars in the Dipper—the ones that make up the front of the bowl—line up as "pointer stars" to the North Star. (The North Star is the last star in the handle of the Little Dipper, another star picture name for the Little Bear.) As its name suggests, the North Star always points north.

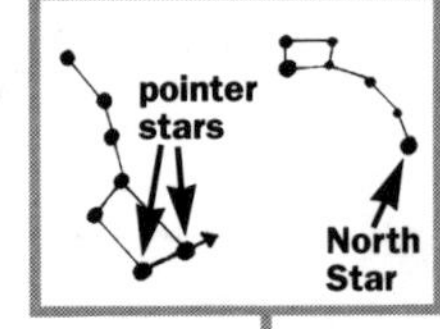

Before the Civil War, escaping slaves used the pointer stars and North Star to find their way north to freedom. They memorized the song "Follow the Drinking Gourd" (the drinking gourd meaning the Big Dipper), which included cleverly encoded additional directions they would need to complete their dangerous journey.

Rather than color the See-Through, have students use a pencil point to poke a tiny hole through each star. Then make a planetarium in your classroom. Dim the lights and let students take turns holding their See-Throughs in front of a wall and shining a flashlight through the holes. The constellations will appear on the wall.

Star Stories (page 47)

Students make up their own myth about the Big Dipper constellation.

After students have written their own star stories, invite them to take home their See-Through and, with an adult, go outside on a clear night to look for these constellations in the sky.

Science See-Throughs That Teach, page 45 Scholastic Professional Books

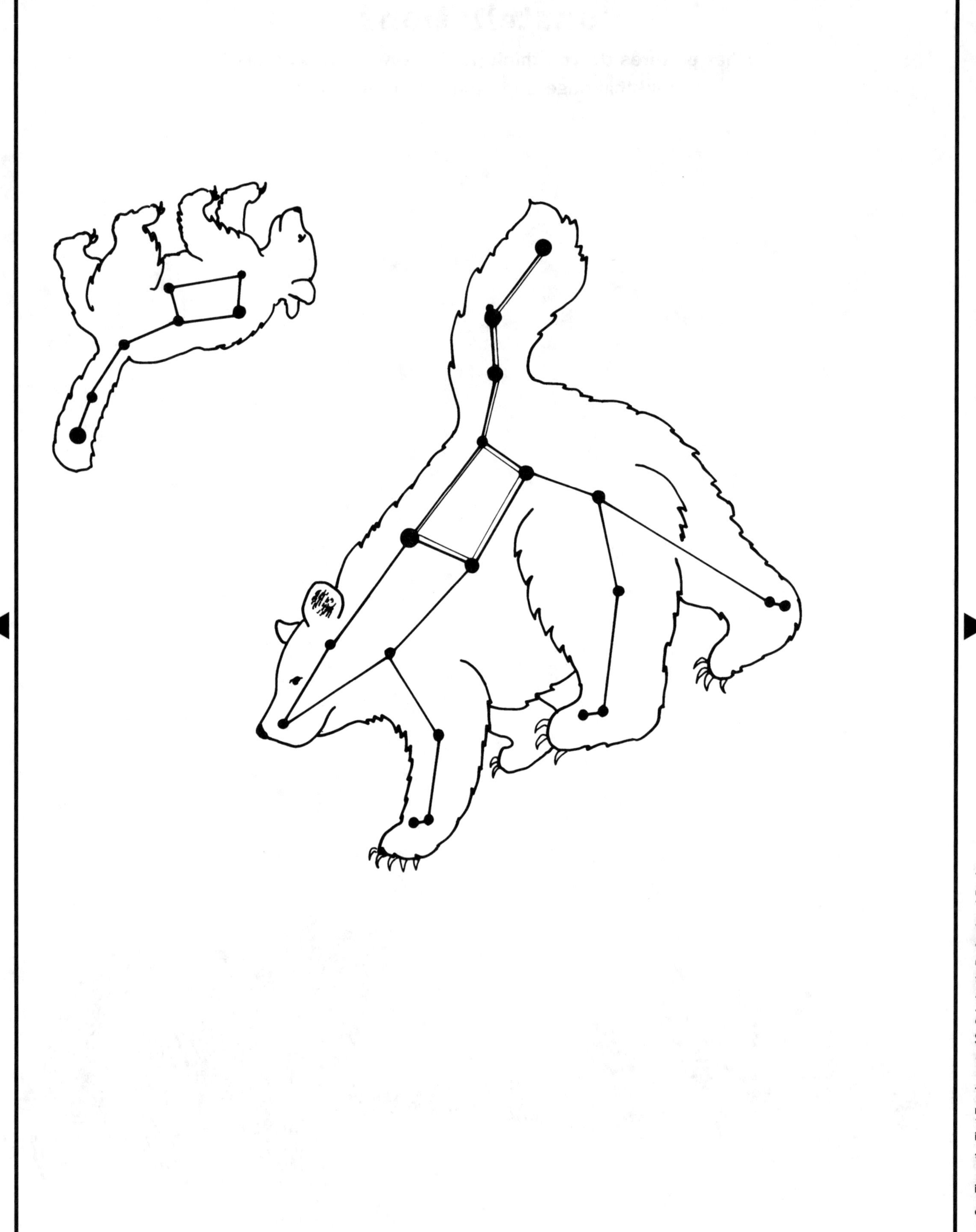

Name: ____________________

Star Stories

These 7 stars make up the Big Dipper. Take a look at the different pictures people have seen in these stars.

People in India saw 7 wise men.

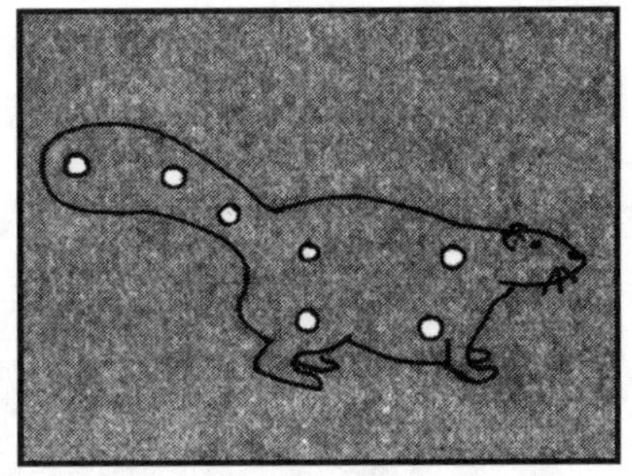

The Sioux (SOO) Indians saw a skunk.

The Chinese saw an emperor on a chariot.

Now make up your own star story! First, draw a picture of the picture you see using the stars below. Then, write a story about your constellation on the back of this page. Tell how it got into the sky.

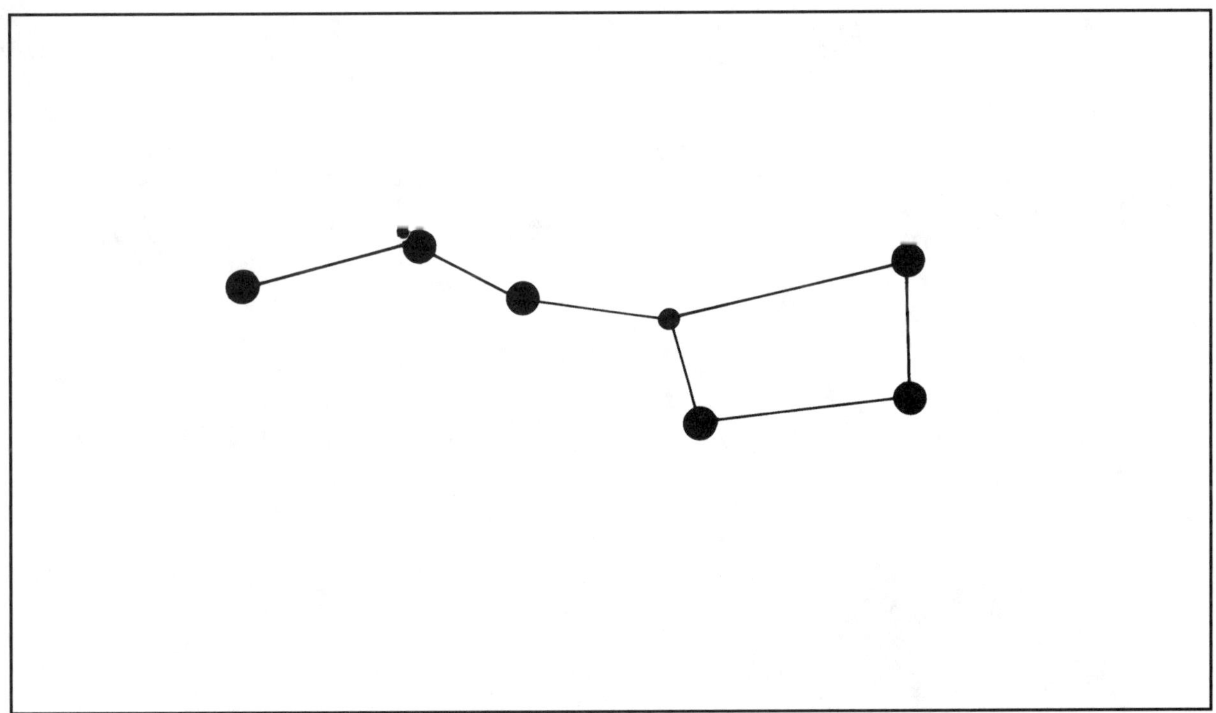

My constellation's name: ____________________

Notes